VEGETABLES

by Robin Nelson

first step non-fiction

Lerner

Lerner Books · London · New York · Minneapolis

We need to eat many
different foods to stay **healthy**.

We need to eat foods in
the **vegetable** group.

Vegetables are parts of plants.

Vegetables give us **vitamins** and **minerals**.

Vegetables help our bodies
heal.

Vegetables help our eyes
stay healthy.

We need three **servings** of vegetables each day.

We can eat lettuce.

We can eat sweetcorn.

We can eat carrots.

We can eat peas.

We can eat broccoli.

We can eat sweet potatoes.

We can eat green beans.

We can eat peppers.

Vegetables keep me healthy.

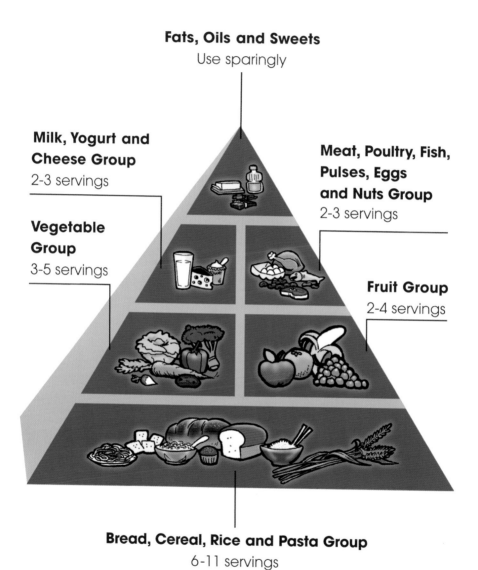

Fats, Oils and Sweets
Use sparingly

**Milk, Yogurt and
Cheese Group**
2-3 servings

**Meat, Poultry, Fish,
Pulses, Eggs
and Nuts Group**
2-3 servings

**Vegetable
Group**
3-5 servings

Fruit Group
2-4 servings

Bread, Cereal, Rice and Pasta Group
6-11 servings

18

Vegetable Group

The food pyramid shows us how many servings of different foods we should eat every day. The vegetable group is on the second level of the food pyramid. You need 3–5 servings of vegetables every day. You could eat a handful of lettuce or 2 stems of broccoli. You could drink a glass of vegetable juice. Vegetables give you vitamins to help you see. Eating vegetables helps our bodies to heal cuts and bruises.

Vegetable Facts

 Vegetables come from plants. The part you eat could be the stem, flower, leaf, seed or root of the plant.

 The onion is named after the Latin word meaning large pearl.

 Most pea pods contain an average of 8 peas.

Carrots contain a lot of vitamin A. Vitamin A helps your eyes to see, especially at night.

 Darker green lettuce leaves are better for you than lighter green leaves.

 Frozen chips were first sold in Britain in the 1960s.

Glossary

 healthy – not sick; well

 minerals – parts of food that keep your blood, bones and teeth healthy

 servings – amounts of food

 vegetable – a part of a plant that you can eat

 vitamins – parts of food that keep your body healthy

Index

This book was first published in the United States of America in 2003.

First published in the United Kingdom in 2008 by
Lerner Books,
Dalton House,
60 Windsor Avenue,
London SW19 2RR

Website address: www.lernerbooks.co.uk

This edition was updated and edited for UK publication by Discovery Books Ltd., Unit 3, 37 Watling Street, Leintwardine, Shropshire SY7 0LW

Words in **bold** type are explained in the glossary on page 22.

British Library Cataloguing in Publication Data

Nelson, Robin, 1971-
   Vegetables. - (First step non-fiction. Food groups)
   1. Vegetables in human nutrition - Juvenile literature
   2. Vegetables - Juvenile literature
   I. Title
   641.3'5

   ISBN-13: 978 1 58013 392 0

The photographs in this book have been reproduced through the courtesy of: © Todd Strand/ Independent Picture Service, front cover, pp 5, 6, 7, 8, 9, 11, 12, 13, 15, 16, 17, 22 (top, second from top, middle, bottom). © PhotoDisc/Getty Images, p 2. © Xavi Arnau/Istockphoto.com, p 3, USDA, pp 4, 10, 22 (second from bottom); © Andrew Cribb/Istockphoto, p 14.

The illustration on page 18 is by Bill Hauser/Independent Picture Service.

Printed in China